S0-ARP-554

Saving Salila's Turtle

An Environmental Engineering Story

Written by the Engineering is Elementary Team
Illustrated by Jeannette Martin

Chapter One | A Rainy Day Discovery

Rainclouds hung dark overhead as my father and I walked home along the banks of the *Ganga Ma*—Mother Ganges. The river currents swirled, making choppy waves in the water.

Turning to the river, I asked, "Are you upset today?"

"Salila," said Father, "you talk to the river so often that I think some day the water might actually answer you. Your mother and I named you well."

I smiled. My name, Salila, is the Hindi word for water, and I've always felt a special connection to the river. I looked away from the Ganges as a fat raindrop plopped onto the book I was carrying. It was the beginning of monsoon season. The raindrops fell faster, making dark splotches on my school uniform.

"Come on, Salila, let's go," Father said. I turned to take one last look at the river before hurrying home.

In that moment, something caught my eye. "Wait, Father, look!" I cried as I pointed.

"What is it?" Father asked, scanning the bank of the river.

I reached down and picked up a small *kachua*, a turtle, that was climbing out of the water. I placed it in my hand, sheltering the turtle from the rain, which was coming down even harder now. Through my wet bangs, I looked towards the spot where the turtle had emerged. The water was covered with oil that created slick, shimmering rainbows.

"Salila, we have to go," Father said as raindrops trickled off his nose. "Put the *kachua* back in the water."

"I can't leave the turtle here!" I cried. "The water is dirty! Won't it make the turtle sick?"

"Salila, this is the turtle's home. And it's pouring. Maybe you can come back later," said Father. "Let's go."

Reluctantly I left the turtle where I'd found it, making sure I remembered the exact spot so I could return to look for it later. How had I never noticed that my beautiful Ganges, important to so many animals, was dirty? As the rain poured down around me, all I could think about was making the Ganges clean.

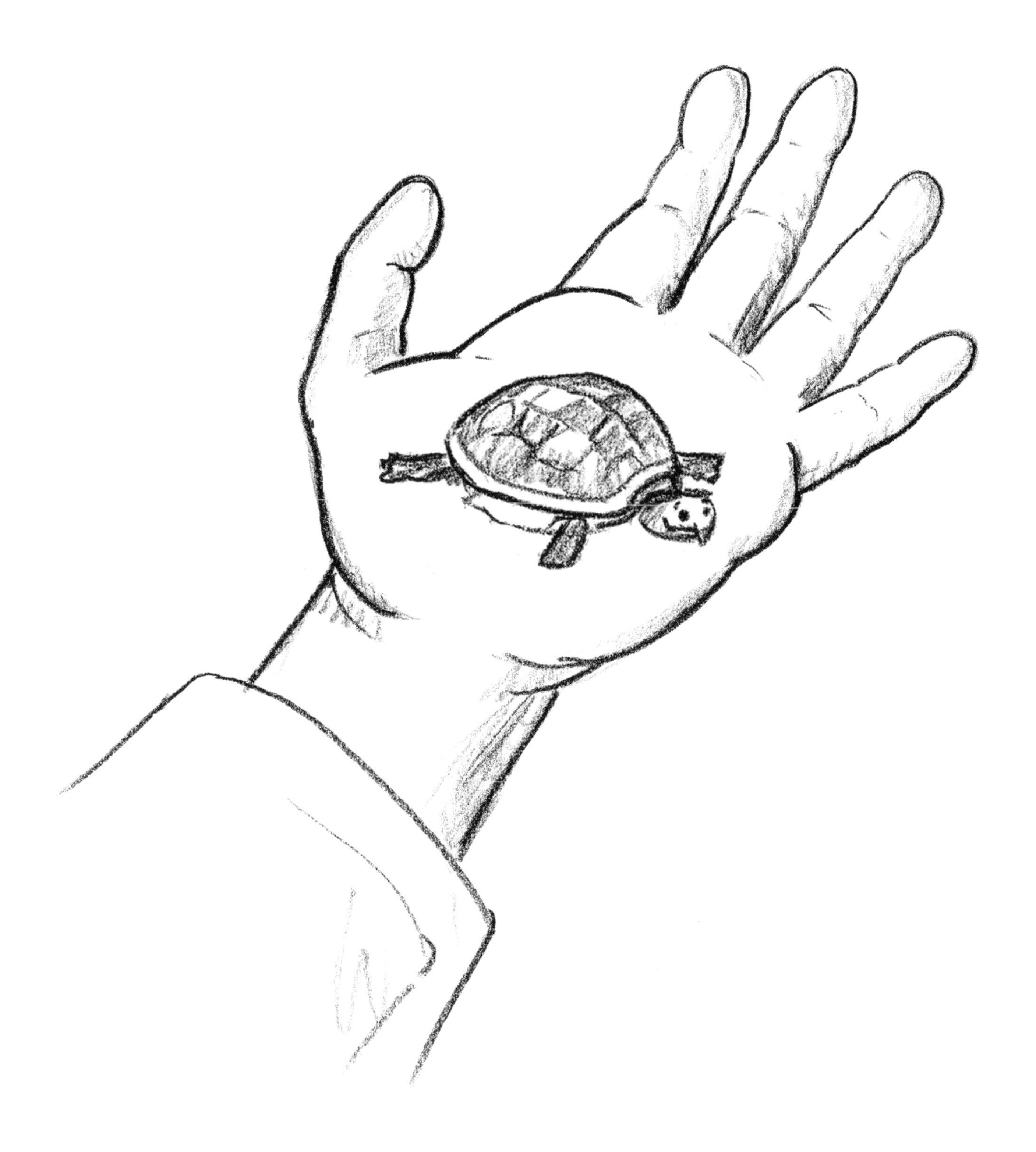

Chapter Two | Pollution and the Water Cycle

"You're soaked!" my mother cried, standing in the doorway. "Don't track that mud through the house."

"Mother, it was horrible," I said as she started to rub my wet hair with a towel. "On the way home from school we found a *kachua* in the Ganges. The water was oily and dirty. I know it's going to make the turtle sick. And what about the other animals living in the water? I need to make the Ganges safe for them!"

Mother looked towards Father and smiled. She shook her head. "It looks like Salila loves animals as much as she loves the water. Okay, Salila," Mother said. "Take a deep breath. We'll figure out a way to help your turtle and the river."

I knew my mother would be able to help. She knows a lot about the water in our country, because she's an environmental engineer. She uses her knowledge of science and math and her creativity to help make the environment cleaner and safer.

"Problems like these are exactly why I became an environmental engineer, Salila," said Mother. "India is a country surrounded by water, and we have many rivers—yet there are still people who don't have access to clean water. We think of the Ganges as sacred because the water allows us to live. But we need to keep the Ganges clean in order for her to continue giving life."

"Father," I said. "I knew we should have rescued the *kachua*!"

Father ruffled my hair. "Oh, Salila. It *is* important to help the environment and the animals who live in it. But

what would we have done with the turtle? How would we have kept it here?"

"Maybe I could keep the turtle in a tank. And I could clean up some water from the Ganges to fill it with! Then the turtle will feel at home. I just need to figure out how to clean the water."

"How about this?" Mother said. "Salila, tomorrow after school I'll take you to the university to show you the work some of my colleagues have done to clean water. Then I think you'll be able to come up with a plan on your own."

"Tomorrow?" I asked. "What should I do to get ready?"

"Well," Mother began, "you can start by asking some good questions. Where does water come from? How can it become polluted—filled with waste that can harm and dirty the environment?"

Father nodded. "Come, Salila," he said. "I'll try to help you answer some of those questions."

As Mother disappeared into her office, I walked over to the sink to get a drink. "There's the answer to the first question Mother asked," I said. "Water comes from the faucet. And before it comes from the faucet, it comes from the river."

"You're right," Father said. "But what about before that? Do you remember learning about the water cycle in school?"

I took a sip of my water, imagining the travels it had taken to get into my glass. "I remember that we learned about water falling from the sky as rain, and water melting from glaciers in the mountains and flowing into streams. And then some of the water from streams and oceans dries up and goes back into the air," I said.

"That's right," Father said. "It's called evaporation."

"So if our water goes through the air and the ground and then into our streams," I began, "then I bet you have to clean all of those areas to really get clean water."

"That's true," Father said.

"Wow," I said. "I guess I never realized that it's important to stop all kinds of pollution so we can make our water cleaner. I always see boats traveling on the river, people cleaning their clothes in the water, and even people taking baths with soap. All of those things must add pollutants. Then there's that factory across the river that sends smoke into the air. Do you think that's bad for the river, too?"

"I bet it is," said Father. "So what do you think we should do?"

"I'm not sure," I said. "It's such a big problem."

"Don't worry, Salila," said Father. "After you go to Mother's office tomorrow, I bet you'll have some new ideas."

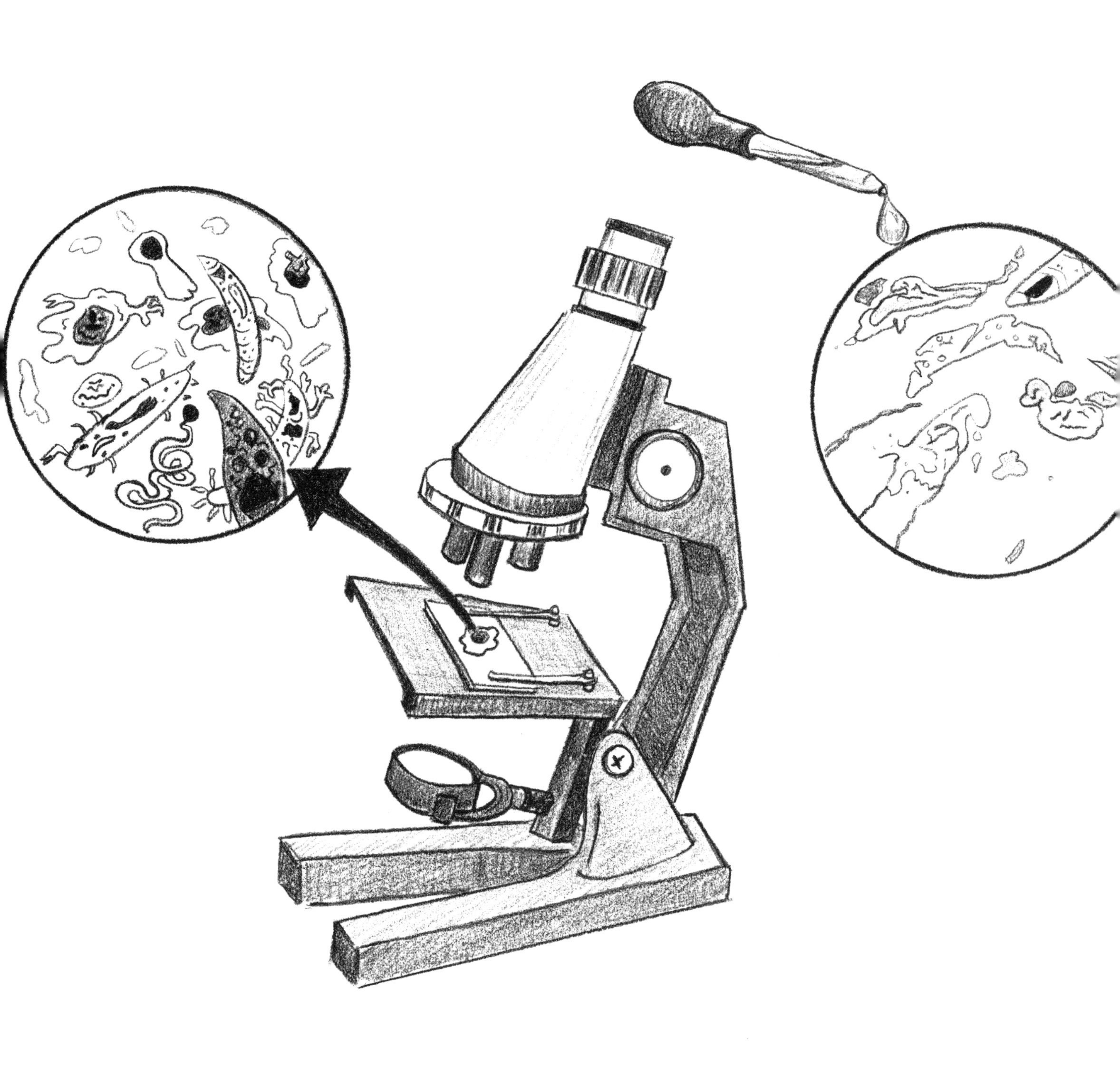

Chapter Three | A Visit to the Lab

The next day after school, I found myself watching strange translucent shapes wriggling and drifting in a drop of water, through a circle of light.

"The water is alive!" I cried, amazed at the tiny creatures I saw through the microscope.

The man beside me, Dr. Gadgil, chuckled. "All different types of microbes live in our water. It's very surprising to see them for the first time."

I turned to Mother, who was sitting next to me. "I wanted you to see the microbes for yourself, Salila," she said. "Those tiny creatures in the water are part of the challenge of keeping water safe and clean."

"That's right," Dr. Gadgil said. "Some of the microbes

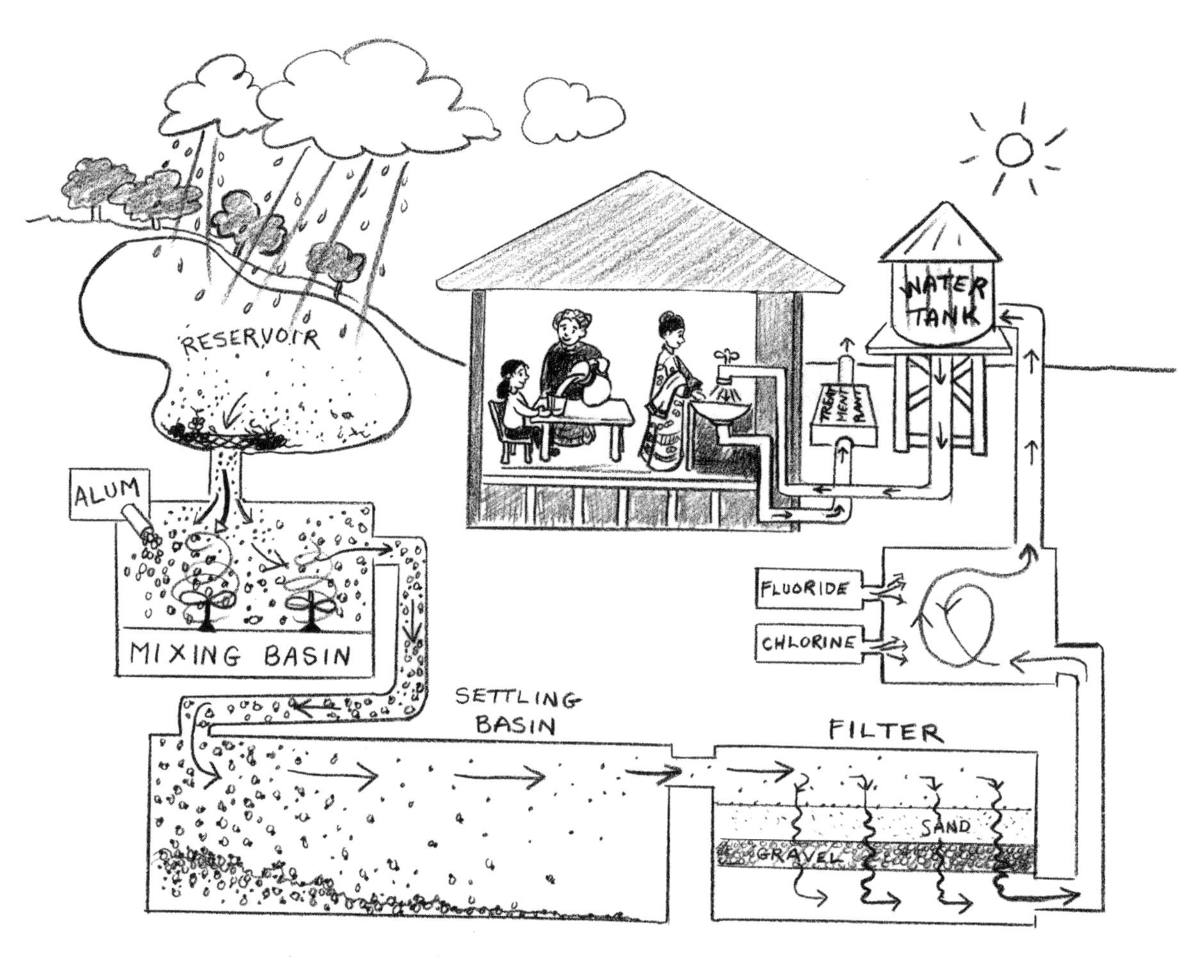

are tiny plants and animals. Some are another type of living thing called bacteria. Not all of them are harmful, but even one harmful microbe could make you sick."

"How can we get these microbes out of the water when we can't even see them without a microscope?" I asked.

"Well, there are a few ways to do it," said Dr. Gadgil. "In many cities, people have their water cleaned at a purification plant before it comes through the pipes to their houses. Dirt, twigs, and microbes are screened out, and then a chemical called chlorine is added to kill any of the microbes that are left."

"But what about people who don't have water brought into their houses, like my grandmother? In her village you have to go to the river to get your water."

"That's a great question. Have you ever seen how your grandmother purifies the water she drinks?" Dr. Gadgil asked.

"I've seen her make a fire to boil her water," I said.

Dr. Gadgil nodded. "Many people in India boil their water to kill the germs. But there isn't always enough wood to burn, so some people drink water without boiling it. That can make them sick. I decided to come up with another way to purify water."

Dr. Gadgil led me over to a white box on Mother's desk. "This is the water purifier I designed to get rid of harmful microbes. I call it the Ultraviolet Waterworks."

I walked around the box and looked at all four of its plain, smooth sides. "How does it work?" I asked.

"It uses light to make the water safer," he said. I frowned, unsure of exactly what Dr. Gadgil meant. "Think of it this way: What happens when you stay in the sun for too long?"

"You can get heatstroke or a sunburn," I said.

Dr. Gadgil nodded. "Exactly. This box has a filter and a special lightbulb in it—an ultraviolet lightbulb. The rays of ultraviolet light are like intense sunshine. They burn the microbes. Then the microbes die and they can't harm people.

That's when the water has been purified. The Waterworks is so small that I am able to bring it to many villages to show people how they can clean their water."

I stared at the box on Mother's desk. *This little machine can purify water?* "I don't know if I can come up with something as fancy as this," I said.

"Technology doesn't always have to be fancy," said Dr. Gadgil. "Technology is really just anything that people create to solve problems. I'm sure you'll come up with some sort of filter that will work well for your turtle."

I was not as sure as Dr. Gadgil. I wondered how that "something" might work. Everything I was learning was interesting, but I still didn't feel any closer to solving the problem of cleaning Ganges water for the *kachua*.

"I still feel stuck," I said to Mother. "I don't know where to start. And if I can't start, then I can't see how I'll solve the problem!"

"Maybe you could use the engineering design process to help you," Mother suggested.

"What's that?" I asked.

"It's a series of steps that I use all the time when I want to solve a problem," she said. "Actually, you have begun it already, even though you didn't know it. You're asking lots of questions."

"I used the engineering design process to help me design the Waterworks," Dr. Gadgil said. "I started by asking lots of questions about water pollution and the ways that people purify their water."

"With all the information you've gathered here today, Salila," said Mother, "you can begin to imagine what your

filter might be like. Then, when you feel like you're ready, you make a plan for your filter—and then create it."

"But what if my filter doesn't work?" I asked.

Mother patted my shoulder. "That's easy," she said.

"You'll go on to the next step—improve. You can just keep improving until you've got something that is the best you can make it."

I nodded. Making a filter might take a lot of thinking and testing, but I had to try to save that turtle. I spent the rest of the day thinking about possibilities.

Chapter Four | Designing a Filter

When I returned home, I started to gather some materials to use for my filter. I found an old shirt in my room—I thought I could try running water through that to strain out big pieces. Then I found some cheesecloth that Mother uses to make *paneer*. Just as I was about to run to the river to fill a bottle with water, I bumped into Father.

"Where are you off to?" Father asked. I told him about the filter materials that I had already gathered.

"While you're at the Ganges," Father said, "why don't you gather some sand from the riverbank? We could try using that to filter the water, too."

"Sand?" I asked. "Won't sand make the water dirty?"

"I think it's worth testing," Father said. "When you

pack sand together there are little spaces between the grains. It could trap some of the dirt and oil in the water and screen it out."

"Screen!" I cried. "I have some mosquito netting in my room that's a really fine screen. We can try that too."

"Now you're thinking!" said Father. "When you get back from the river, I'll help you test these things out."

As I walked back home from the river, I lifted up the bottle of water. Sunlight streamed through it, making the water glow like amber. I remembered the microbes we had seen through the microscope in Mother's lab and wondered how many little creatures were inside the bottle.

Father helped me set up a few different filters. We decided to test the mosquito netting first.

"Ready, set, pour!" I called.

I watched the water pour through the filter and settle in the bottom of the cup. "The water still looks kind of brown," I said, turning to Father.

"This is just the first step," said Father. "Remember what your mother said about the engineering design process?"

I nodded. "I guess we should test everything before I come up with a plan."

After Father and I had tested all of the materials on

their own, I started experimenting with mixing materials together. I layered sand between fabric and cheesecloth. I tried running water through pebbles and cotton. When we thought we had improved our filter so that it was the best one possible, I began filling a tank with the clean water. Now I had to find the turtle!

Chapter Five

Hope for the Ganges

The next day I sat on the bank of the Ganges, thinking about everything I had learned in the past few days. Eventually, the *kachua* appeared, climbing out of the water exactly where I'd seen her during the rainstorm. I carefully lifted her up and looked out at the river.

There were people using the water to wash and boats traveling. As usual, the factory across the way blew billows of gray smoke across the sky.

But today there was also another person using the river. It was a woman, performing a ritual cleansing in the water. Dressed in an elegant, colorful *sari*, the woman scooped water into a jug and let the water pour out of the container, falling in an arc back into the Ganges.

It's so beautiful, I thought. *I wish more people were here to see this.* With that simple thought, I knew how I would help not just this one turtle, but the whole Ganges.

When I got home, Mother was working in the kitchen. I placed the turtle in her new tank and smiled broadly.

"Salila," said Mother, "You did a great job. I'm proud of you."

"Thanks, Mother," I said. "But I'm not done yet. Really, if we want the river to be clean, we need to stop pollution in lots of different areas, not just in the river. I'm going to talk to my friends about everything I learned from you and Dr. Gadgil. I want to show them what I did for my new turtle. And then I'm going to ask them to help me clean up the Ganges."

"That's a great idea!" said Mother. "If your friends start to take care of the environment, then it will spread to their friends, and their friends will tell even more people."

"And maybe someday," I said, "if we all work really hard, I can bring the turtle back to the river so she can live there again. We could make the Ganges pollution-free for animals and for people!"

Try It!

Design a Water Filter

You can design a water filter, just like Salila. Imagine that there is a pond by your house and you would like to filter some of its water. Your goal is to make the water clearer and cleaner than it was before it was filtered!

Materials

- ☐ Two empty soda bottles with caps
- ☐ Clear packing tape
- ☐ Clear plastic cups or a basin
- ☐ Loose tea
- ☐ Potting soil
- ☐ Coffee filters
- ☐ Cheesecloth
- ☐ Sand or gravel

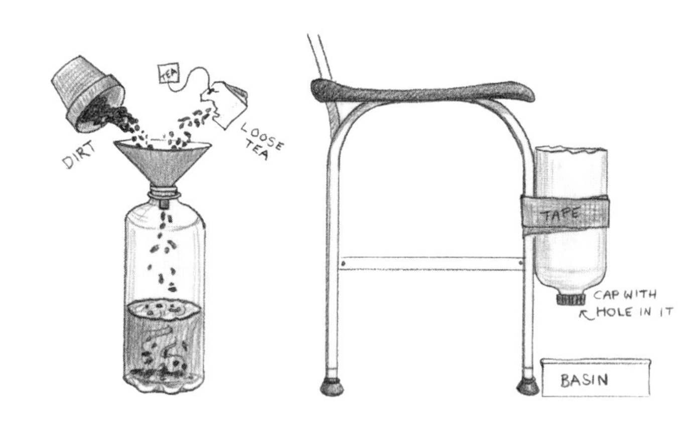

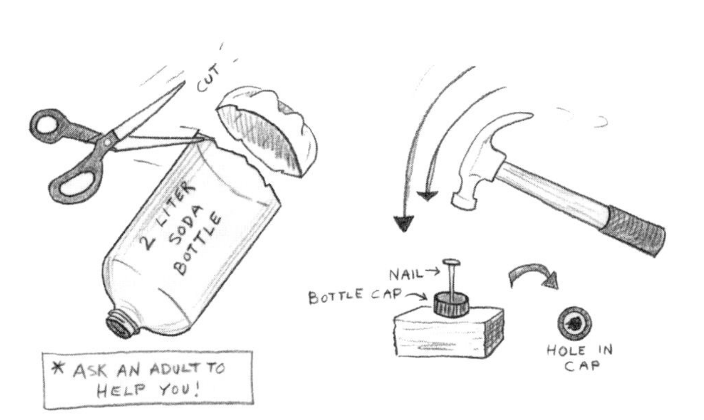

Set Up

Ask an adult to help you cut the bottom off an empty soda bottle and make a small hole in the cap with a hammer and a nail or with a drill. Put the cap back on the bottle and tape it to a chair leg. You can design your filter to fit in this bottle.
To make your polluted water, put a few tablespoons of loose tea in a whole, empty soda bottle along with hot water. Put the cap on the bottle and shake it up. Let it sit until the water is dark brown. Add a cup of potting soil and shake it up.

Test the Materials

Fill the cut soda bottle with materials that you think might help filter the pollutants out of your water. Test each material by itself. How does the water look after you run it through each material once? Twice? Does the water flow through each material at the same speed or at different speeds?

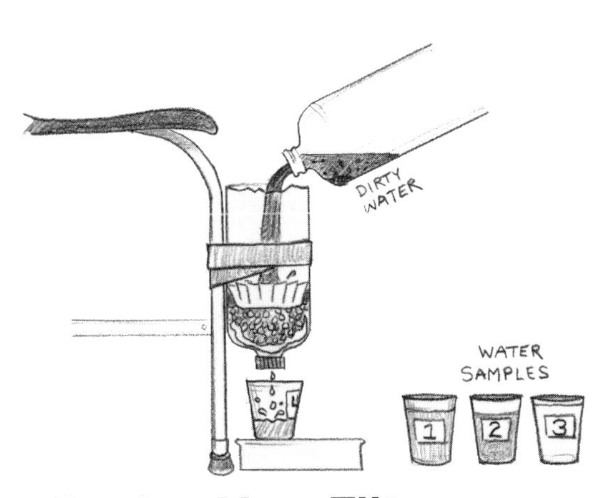

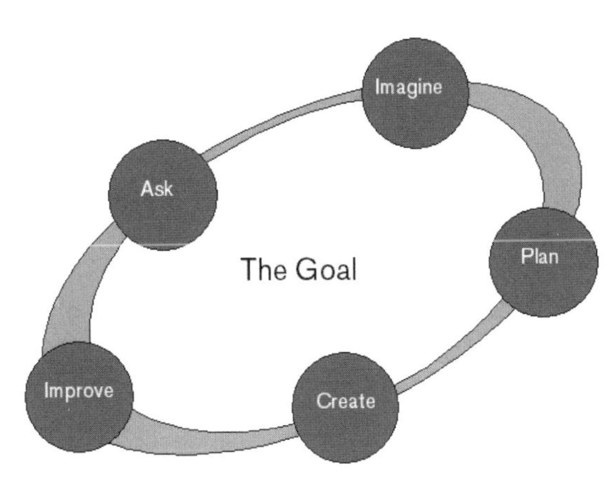

Design Your Filter

Use what you learned in your tests to make a plan for your final filter system. Think about how much of each material you will need and how you will assemble them so that your design is the best it can be.

Improve Your Filter

Use the engineering design process to improve your filter. Can you think of more filter materials that you'd like to test? Will your filter help to clean a different type of polluted water?

See What Others Have Done

See what other kids have done at http://www.mos.org/eie/tryit. What did you try? You can submit your solutions and pictures to our website, and maybe we'll post your submission!

Glossary

Bacteria: Tiny, one-celled organisms that are too small to see without a microscope. Bacteria are not animals or plants.

Chlorine: A chemical that is used to kill microbes in water.

Engineer: A person who uses his or her creativity and understanding of mathematics and science to design things that solve problems.

Engineering Design Process: The steps that engineers use to design something to solve a problem.

Environment: The natural world in which we live.

Environmental Engineering: The branch of engineering concerned with solving problems with air, water, soil, and the natural environment.

Evaporation: The process of changing from a liquid into a gas.

Filter: Something that separates impurities from a liquid or a gas.

Glacier: A huge, slow-moving body of ice.

Kachua: The Hindi word for turtle. Pronounced *ka-chew-ah*.

Microbes: Living things that are too small to see without a microscope. Bacteria are one type of microbe.

Monsoon: A wind that brings summer rain to South Asia.

Paneer: A type of cheese common in South Asian cuisine. Pronounced *pan-ear*.

Pollution: Human-made waste that dirties the environment.

Sari: A piece of cloth that is draped over the body and worn as a garment by Hindu women. Pronounced *sah-ree*.

Technology: Any thing or process that people create and use to solve a problem.

Ultraviolet Light: A type of light not visible to the human eye. It can burn living things.

Water Purification: The process of removing from water all of the pollution and other things that are not water.

Water Vapor: Water that has evaporated and is in the form of a gas.